SPOTLIGHT ON OUR FUTURE

CLIMATE CHANGE AND OUR EARTH

KATHY FURGANG

PowerKiDS press™

NEW YORK

Published in 2022 by The Rosen Publishing Group, Inc.
29 East 21st Street, New York, NY 10010

Copyright © 2022 by The Rosen Publishing Group, Inc.

All rights reserved. No part of this book may be reproduced in any form without permission in writing from the publisher, except by a reviewer.

First Edition

Editor: Theresa Emminizer
Book Design: Michael Flynn

Photo Credits: Cover Pavel_Klimenko/Shutterstock.com; (series globe background) photastic/Shutterstock.com; p. 5 jo Crebbin/Shutterstock.com; p. 6 Gino Santa Maria/Shutterstock.com; p. 7 https://commons.wikimedia.org/wiki/File:Sunset_from_the_ISS.JPG; p. 9 rybarmarekk/Shutterstock.com; p. 10 lavizzara/Shutterstock.com; p. 11 -/AFP/Getty Images; p. 12 Siberian Art/Shutterstock.com; p. 13 Stock Montage/Archive Photos/Getty Images; p. 15 (John Tyndall) UniversalImagesGroup/Getty Images; p. 15 (Guy Stewart Callendar) https://en.wikipedia.org/wiki/Guy_Stewart_Callendar#/media/File:GSCallendar1934.jpg; p. 17 https://commons.wikimedia.org/wiki/File:Charles_David_Keeling_%26_GW_Bush_2001.jpg; p. 18 Dennis Cook/AP Images; pp. 19, 24 NurPhoto/Getty Images; p. 20 nicostock/Shutterstock.com; p. 21 Spencer Platt/Getty Images; p. 23 Sean Gallup/Getty Images; p. 25 Mizzick/Shutterstock.com; p. 26 John D Sirlin/Shutterstock.com; p. 27 David McNew/Getty Images; p. 29 kwest/Shutterstock.com.

Cataloging-in-Publication Data

Names: Furgang, Kathy.
Title: Climate change and our Earth / Kathy Furgang.
Description: New York : PowerKids Press, 2022. | Series: Spotlight on our future | Includes glossary and index.
Identifiers: ISBN 9781725323797 (pbk.) | ISBN 9781725323827 (library bound) | ISBN 9781725323803 (6 pack)
Subjects: LCSH: Climatic changes--Juvenile literature. | Climatic changes--Effect of human beings on--Juvenile literature. | Global warming--Juvenile literature. | Nature--Effect of human beings on--Juvenile literature.
Classification: LCC QC903.15 F87 2022 | DDC 363.738'74--dc23

Manufactured in the United States of America

Some of the images in this book illustrate individuals who are models. The depictions do not imply actual situations or events.

CPSIA Compliance Information: Batch #CSPK22. For further information contact Rosen Publishing, New York, New York at 1-800-237-9932.

CONTENTS

RISING POPULATIONS, SUFFERING SPECIES. 4
EARTH'S CHANGING WEATHER. 6
CAUSE AND EFFECT 8
EARTH IN DANGER 10
THE GREENHOUSE EFFECT 12
CLIMATE CHANGE FROM BUSINESS 14
CARBON DIOXIDE LEVELS 16
IT'S GETTING WARMER. 18
THE WORLD WORKS TOGETHER. 20
A NEW REPORT. 22
CLIMATE RISKS 24
CHANGING WEATHER PROBLEMS 26
NEW IDEAS, NEW ENERGY 28
STANDING FOR CHANGE 30
GLOSSARY 31
INDEX 32
PRIMARY SOURCE LIST 32
WEBSITES. 32

CHAPTER ONE

RISING POPULATIONS, SUFFERING SPECIES

There are almost 8 billion humans on Earth. That's a lot of people! So many, in fact, that our activities are creating changes to the planet's climate, or average weather conditions. Climate change affects all living things on Earth.

As the human population grows, people move into new areas and change the environment, or natural world. As a result, many animal **species** are suffering. Every year, about 30,000 species are being pushed to extinction. Extinction is when all members of a species are dead.

Scientists are working hard to solve these problems. But it's important for other people to help, too. Many **activists** are fighting to create a better future for our world. By making small changes, you can help slow the harmful effects of climate change and save disappearing species.

Climate change is causing sea levels to rise, which results in flooding. Flooding harms animals, such as the Florida panther, which lives in low-lying areas.

CHAPTER TWO

EARTH'S CHANGING WEATHER

Earth's atmosphere is made up of a mixture of gases. It's about 78 percent nitrogen, 21 percent oxygen, and less than 1 percent other gases such as carbon dioxide (CO_2). Human activities are changing the mixture of gases in the atmosphere. This is also changing the climate.

This is a picture of Earth's atmosphere as it looks from space.

Fossil fuels are fuels—such as coal, oil, or natural gas—that formed in the earth from dead plants or animals. Over the last 150 years, humans have burned fossil fuels to create the energy needed to make goods and power machines. This puts **greenhouse gases** such as carbon dioxide into the air.

These gases trap heat, warming Earth. This is called the greenhouse effect. To keep the planet's gases in balance, people are trying to reduce, or lessen, the use of fossil fuels.

CHAPTER THREE
CAUSE AND EFFECT

The more fossil fuels humans burn, the more the atmospheric balance is altered. Scientists believe that releasing more greenhouse gases could have serious consequences for Earth's climate. If we don't curb greenhouse gases soon, it may be too late.

Increased heat could cause permafrost to melt. Permafrost is a soil layer that's always frozen in very cold parts of the world. If it thaws, or warms up, even more greenhouse gases could be released, or let out. This would add more heat-trapping gases to the atmosphere and raise Earth's temperature even further.

Over the last 200 years, the human population has risen from 1 billion people to almost 8 billion people. Countries such as the United States release a lot of greenhouse gases. Climate changes cause droughts, or dry periods, increased heat, and bad weather.

As climate change causes ice to melt, trapped greenhouse gases such as methane are released.

CHAPTER FOUR

EARTH IN DANGER

People are changing the natural cycles that make Earth livable. The water cycle has been greatly affected by climate change. As Earth gets warmer, more water **evaporates** into the air. More water in the air causes greater precipitation, or rainfall and snowfall. This can cause heavy flooding and storms.

In 2017, Hurricane Harvey hit southeastern Texas. The storm ruined many buildings and claimed 88 lives. The following year, Hurricane Florence hit North Carolina, wrecking cities and killing about 40 people.

HURRICANE HARVEY, 2017

As rainfall increases around the world, so does flooding. This flood in India happened in 2018.

Climate change is also causing longer droughts. Droughts can cause erosion, or the wearing away of the land. They can also cause crops to fail. California had its longest drought so far from 2012 to 2017.

Heavier rains and longer dry periods have harmful effects on communities and environments around the world.

CHAPTER FIVE

THE GREENHOUSE EFFECT

Earth's atmosphere naturally traps some of the heat from the sun. However, climate change is causing the atmosphere to trap too much heat. In the 1800s, scientists did experiments to see if gases created by human activities could heat Earth.

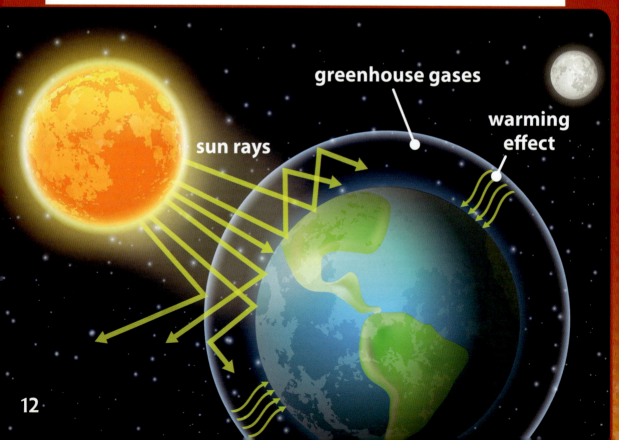

Joseph Fourier's greenhouse box experiment helped people understand how climate change happens.

In 1824, a French scientist named Joseph Fourier studied how Earth stays warm. He found that the atmosphere traps heat from the sun and **insulates** Earth. Without the atmosphere, Earth would be about 60°Fahrenheit (33°Celsius) cooler.

Fourier experimented with a box and a glass cover to represent Earth and the atmosphere. When sunlight entered the box, the glass trapped heat. The temperature inside the box got warmer. Even though the experiment was simple, Fourier's ideas shaped how people understand the effects of greenhouse gases.

CHAPTER SIX

CLIMATE CHANGE FROM BUSINESS

During the late 18th century, the Industrial Revolution began. This was a period of economic change in which machines were increasingly used to create goods. Many machines were powered by steam. People burned coal to run the steam engines, releasing large amounts of harmful greenhouse gases into the air. The Industrial Revolution is often seen as the beginning of major climate change.

In the 1860s, scientist John Tyndall studied greenhouse gases and climate change. He learned that gases from burning coal particularly absorbed heat. These materials include **hydrocarbons**, carbon dioxide, and methane. He learned that carbon dioxide absorbs the sun's heat like a sponge.

In the 1930s, engineer Guy Stewart Callendar observed that the northern Atlantic region and the United States had warmed a good deal after the Industrial Revolution.

JOHN TYNDALL

GUY STEWART CALLENDAR

British engineer Guy Stewart Callendar was one of the first people to observe climate change.

CHAPTER SEVEN

CARBON DIOXIDE LEVELS

In 1958, an American scientist named Dr. Charles Keeling made a large contribution to the study of climate change. While working at an observatory in Hawaii, he found a way to record carbon dioxide levels. Using this information, Keeling made a graph showing how carbon dioxide levels changed over time. This became known as the Keeling Curve. From that time to present day, carbon dioxide has continued to rise. Keeling's study is the longest measure of carbon dioxide levels over time so far.

As human populations continue to rise, people burn more fossil fuels and more carbon dioxide is released. The Keeling Curve shows a steady increase of carbon dioxide in the atmosphere. Scientists can also compare this information with the amounts of carbon dioxide trapped in Earth's ice during earlier times.

In 2002, Dr. Charles Keeling received the Medal of Science from President George W. Bush.

17

CHAPTER EIGHT

IT'S GETTING WARMER

The summer of 1988 was the hottest summer recorded in the United States. The heat caused fires and droughts. Many people could feel the change.

In 1988, NASA scientist Dr. James Hansen spoke to the United States Congress about climate change. He used special dice to show Congress how likely warmer-than-average summers were becoming because of CO_2 **emissions** and climate change. Hansen said he was "99 percent sure" that humans caused climate change. His talk had a huge effect on opinions about the problem.

NASA scientist Dr. James Hansen spoke to the United States Senate about climate change in 1989.

AL GORE

In 2000, Vice President Al Gore ran for president. He spoke about climate change during his campaign. He didn't win the election, but he continued to talk about climate change. In 2007, he was part of a group that won a Nobel Prize for work on climate change.

CHAPTER NINE

THE WORLD WORKS TOGETHER

In 1989, the United Nations started working on studying climate change. It formed a panel of people and scientists from many countries to learn more about climate change. The panel worked to gather information about climate change, its effects, and how people might need to adjust to these changes.

U.S. Secretary of State John Kerry signed the Paris Climate Agreement in 2016. He's shown holding his granddaughter.

In 2015, many countries came together to create the Paris Climate Agreement. Its goal is to stop the world's temperature from rising more than 3.6°Fahrenheit (2°Celsius) above the temperature before the Industrial Revolution. The United States entered this agreement under President Barack Obama.

Working together is important if we hope to keep our future safe. Rising temperatures cause heat waves, bad storms, long droughts, and higher sea levels. For a better tomorrow, we need to take action today.

CHAPTER TEN

A NEW REPORT

In 2018, the United Nations panel released a new report on climate change. It said that to prevent deadly weather events, global warming must be limited to a rise of 2.7°F (1.5°C). It warned that higher temperatures could lead to major environmental changes, such as the disappearance of coral reefs and more ice melting in the Arctic, causing sea levels to rise. The report said there were only 12 years left to make the changes needed to avoid these **disasters**.

However, U.S. President Donald Trump took steps to withdraw the United States from the Paris Climate Agreement. He said he felt its rules were unfair to the United States. Many people disagreed with his decision. Following the election of President Joe Biden in 2020, the United States rejoined the Paris Climate Agreement in February 2021. Many countries around the world cheered the decision.

Many people protested President Trump's decision to take the United States out of the Paris Agreement.

CHAPTER ELEVEN

CLIMATE RISKS

The effects of climate change can already be seen around the world. The Arctic region is one of the areas affected the most. Arctic sea ice changes every season. Pictures from space show that the amount of Arctic ice has gotten much smaller since 1979.

In 2019, a heavy storm in São Paulo, Brazil, caused a lot of damage.

Glaciers in Iceland and other areas have also changed. One glacier covered about 6.2 square miles (16.1 sq km) in 1890. By 2019, the glacier had shrunk so much that it was said to be "dead." People in Iceland held a funeral to call attention to the issue.

Melting water from glaciers is changing the oceans. Sea levels are rising, and ocean currents are slowing. Slowing currents may change the weather around the world. Some storms could be worse. The Northern Hemisphere could also become colder.

CHAPTER TWELVE

CHANGING WEATHER PROBLEMS

Because of climate change, there are now more wildfires in parts of the world. Hotter temperatures cause snow to melt faster and dry out soil. When this happens, plants can catch fire more easily.

When the weather changes, some areas get less rain. Wildfire season is many months longer than it was just a decade ago. In 2018, a big wildfire swept through the town of Paradise, California, destroying most of the town and killing 85 people.

Firefighters in California work to control a wildfire in 2017.

Climate change is causing land to turn into deserts. This is called desertification. Larger deserts are causing bigger dust storms, which can cause health problems in people who live nearby.

Climate issues are forcing people to move. Thousands of people in Africa, South Asia, and Latin America have become **climate refugees**.

CHAPTER THIRTEEN
NEW IDEAS, NEW ENERGY

The number of people on Earth keeps growing. There may be 11.2 billion people by 2100. Countries are working to address climate change issues in new ways, including new **technologies**.

One idea is to use technology to capture, or store, harmful greenhouse gases. Stored greenhouse gases can't make the planet warmer. In 2019, scientists in Iceland first managed to store carbon dioxide underground.

The wind, the sun, and ocean waves are renewable sources of energy. This mean they won't run out. When people use renewable energy sources, they burn fewer fossil fuels. Renewable resources don't release greenhouse gases. In 2017, solar power made up 10 percent of the energy produced in the United States. In 2019, Ireland used ocean waves to create electricity. Some countries are encouraging companies to use renewable resources.

HOW YOU CAN HELP

INDIVIDUAL ACTIONS	IMPACT
Shorten showers by five minutes for one month	Save around 3 trillion gallons (11.3 trillion L) of water
Eat less meat	Save around 25 percent of individual carbon emissions
Line-dry clothes instead of using the dryer once a week	Save the equivalent energy of turning on 225 light bulbs for an hour
Vote for government leaders who have plans to curb climate change	Have a part in shaping your future and Earth's future too!

CHAPTER FOURTEEN

STANDING FOR CHANGE

In 2018, a 15-year-old Swedish girl named Greta Thunberg decided to skip school to help the climate. She went to the Swedish **parliament** building to protest the worldwide climate crisis. Thunberg started the Fridays for Future movement to have students protest every Friday. The movement wants governments to take more action to fix the climate crisis. About 100,000 students from around the world now take part in this movement.

Xiuhtezcatl Martinez started speaking out about climate change when he was only six years old. Now 18, Martinez is the youth director of Earth Guardians, a program devoted to getting teens involved in creating a better future for Earth.

Thunberg and Martinez have shown that the actions of young people can help create a more **sustainable** world.

GLOSSARY

activist (AK-tih-vist) Someone who acts strongly in support of or against an issue.

climate refugee (KLY-muht REH-fyoo-jee) Someone who has to leave an area because of climate-linked issues.

disaster (dih-ZAS-tuhr) Something that happens suddenly and causes much suffering and loss for many people.

emission (ee-MIH-shuhn) Something that is given off, or the act of producing that thing.

evaporate (ih-VA-puh-rayt) To change from a liquid into a gas, or vapor.

greenhouse gas (GREEN-howz GASS) A gas in the atmosphere that traps energy from the sun.

hydrocarbon (HY-droh-kahr-buhn) A substance that contains only carbon and hydrogen.

insulate (IHN-suh-layt) To add material to something to stop heat, electricity, or sound from going into or out of it.

parliament (PAHR-luh-muhnt) A lawmaking body.

species (SPEE-sheez) A group of plants or animals that are all the same kind.

sustainable (suh-STAY-nuh-buhl) Able to last a long time.

technology (tek-NAH-luh-jee) A method that uses science to solve problems and the tools used to solve those problems.

INDEX

A
Africa, 27
atmosphere, 6, 7, 8, 12, 13, 16

C
Callendar, Guy Stewart, 14, 15
carbon dioxide, 6, 7, 14, 16, 18, 28
coal, 7, 14

D
desertification, 27
drought, 8, 11, 21

E
Earth Guardians, 30

F
fossil fuels, 7, 8, 16
Fourier, Joseph, 13

G
Gore, Al, 19
greenhouse gases, 7, 8, 9, 12, 13, 14, 22, 28

H
Hansen, James, 18

I
Iceland, 28
Industrial Revolution, 14, 21
Ireland, 28

K
Keeling, Charles, 16
Keeling Curve, 16

L
Latin America, 27

M
Martinez, Xiuhtezcatl, 30

O
Obama, Barack, 21

P
Paris Agreement, 21, 22

S
South Asia, 27

T
Thunberg, Greta, 30
Trump, Donald, 22
Tyndall, John, 14, 15

U
United Nations (UN), 20, 22
United States, 8, 14, 18, 21, 22, 28

PRIMARY SOURCE LIST

Page 7
The Indian Ocean from the International Space Station. Photograph. June 14, 2010. Now kept at NASA.

Page 15
Guy Stewart Callendar. Photograph. 1934. Now kept at the University of East Anglia Archive.

Page 17
Charles Keeling receives the Medal of Science. Photograph. June 13, 2002. Now kept at the National Science Foundation.

WEBSITES

Due to the changing nature of Internet links, PowerKids Press has developed an online list of websites related to the subject of this book. This site is updated regularly. Please use this link to access the list: www.powerkidslinks.com/SOOF/climatechange